Consider sharing the book with someone who would benefit from it to make the environment and me happy.

WHO IS DR. BEE?

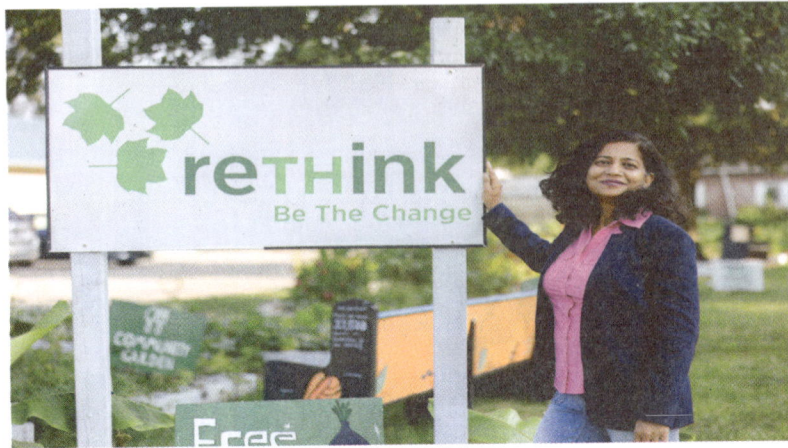

Meet Dr. Shikha Bhattacharyya: A Passionate Environmentalist and Pharmacist.

Originally from India and now residing in the United States, Dr. Shikha Bhattacharyya has a deep passion for creating a cleaner, more sustainable, and healthier world for our children. As a pharmacist and environmental educator, she is dedicated to making a difference in the world.

Aside from her role as a full-time community pharmacist, Dr. Bhattacharyya is also the founder and executive director of reTHink, Inc. This non-profit organization, located in Terre Haute, Indiana, manages four community gardens, a zero-waste community store, and a workspace dedicated to upcycling plastics. Dr. Bhattacharyya has been awarded many awards for her work in the community, including the Green Lights Award by Earth Charter Indiana, Community Educator award by the city of Terre Haute, and Cox Conserves Heroes Groundbreaker award.

FOREWARD

"Children need to learn the importance of our planet. In this book, you'll see all the fun ways you can make a difference. I know this can be done as I, myself, have cleaned up over 30,000 pounds of trash from our waterways and started several recycling programs of my own. These things may not be taught in school as they should be but they are so important for the future of our planet. Every child needs to know there is something they can do and their voice and actions matter! Kids may be a small part of the population but We are 100% of the future and we can help change the world."

-Cash Daniels

The Conservation Kid

theconservationkid.com

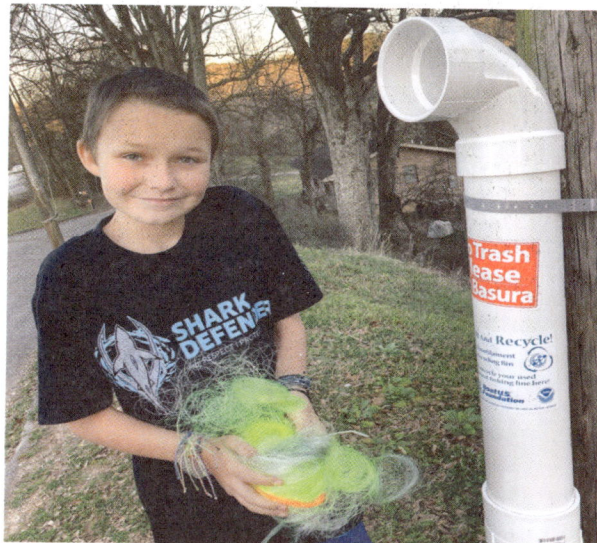

ANSWERS AND FEEDBACK!

Should you require assistance with answers, kindly visit my website at www.busydoctorbee.com or scan the QR Code below. I will provide the answer key and videos designed to assist parents and teachers. You can also reach out to me via email at busydoctorbee@gmail.com. I look forward to receiving your inquiries and feedback!

TABLE OF CONTENTS

INTRODUCTION

> Hi There!
> I'm Dr. Bee.
> I love to talk about our environment.

What's your name? ....

What do you like to talk about?

....

WHAT IS ENVIRONMENT?

What on earth is this "environment" thing anyway?

As far as I know, our environment is the space where we all coexist – it's the air we breathe, the water we drink, and the earth beneath our feet. It also includes animals and plants.

COMPONENTS OF THE ENVIRONMENT

Five essential components of the environment are:

1. Air
2. Water
3. Earth
4. Animals
5. Plants.

WHAT DO WE BREATHE IN?

We breathe in air and use oxygen from it to live. Trees breathe air too! But they take carbon dioxide (CO_2) from the air and give out oxygen (O_2).

Would you like to color these trees?

4

WATER WATER EVERYWHERE!

> 71%
> of the Earth is
> covered with water.
> Unfortunately, not all
> water is drinkable.

Have you ever experienced a shortage of water? Do you want to share your experience?

WE ARE MADE OF WATER!

Did you know that about 60% of the human body is just water?

Guess how long a person can live without water. ....

OUR EARTH

Let's start with the planet Earth. Earth is very special, as it is the only place I know that you and I can live.

Bonus question - Do you know any other place in the universe where humans or other animals might be living?..............

WHAT'S ON THE EARTH?

Earth's got it all! There's the dry parts we call continents, and the wet parts we call oceans. Easy-peasy!

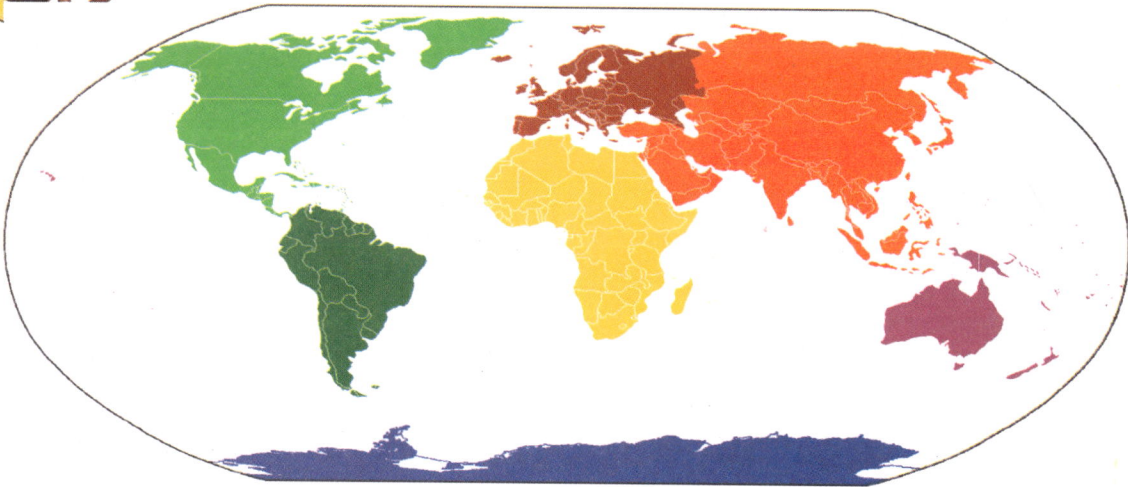

Have you ever seen this map of the world? Can you identify any continents or oceans?

OUR CONTINENTS

How many continents can you count?

Which continent do you live in?...................

OUR OCEANS

How many oceans do you see in this map?

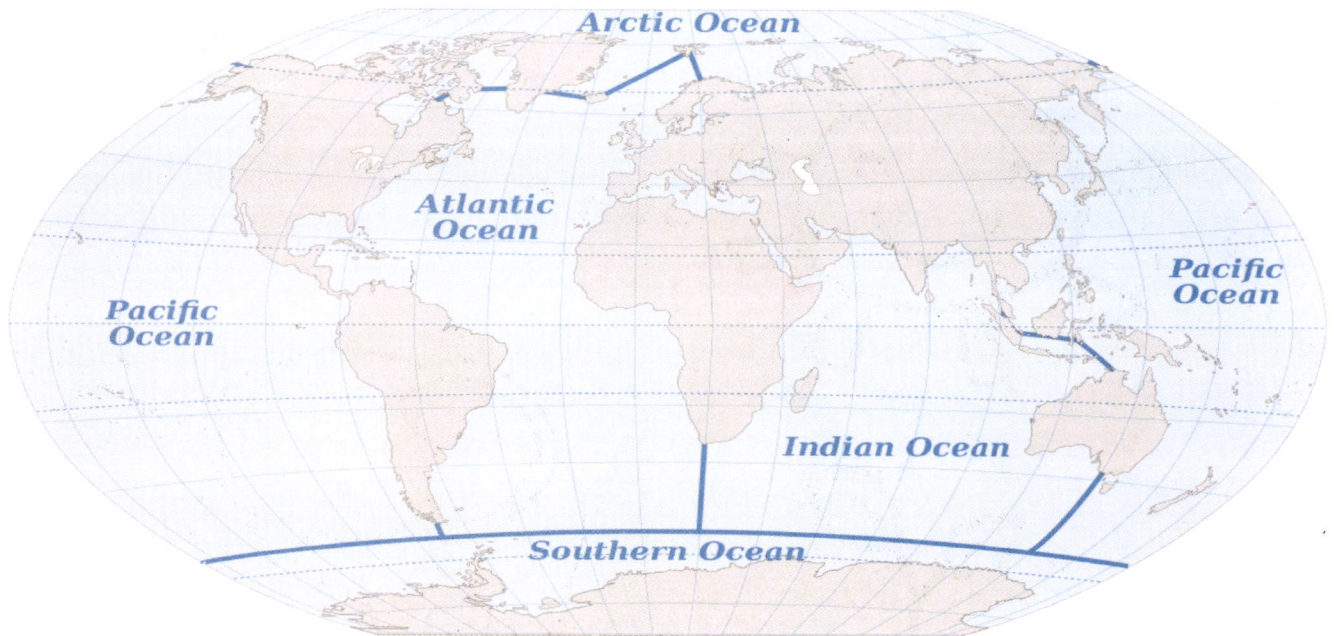

Arctic Ocean

Atlantic Ocean

Pacific Ocean

Pacific Ocean

Indian Ocean

Southern Ocean

Can you circle and read the names of all the oceans?

MY FAVORITE THINGS

I have some favorite things in the environment. Do you want to color them?

Sun

Plant

Elephant

WHY DO WE NEED THE ENVIRONMENT?

Can you imagine living without air, water, and land? I sure can't!

WE ALSO GET FOOD FROM THE ENVIRONMENT

What are 3 of your favorite foods?

1.................

2.................

3.................

My favorite food is mango :).

NATURAL AND MANMADE FOOD

Did you know that the food
that comes straight from
the environment is
generally healthier
than processed food?

Circle and color foods
that come straight from nature

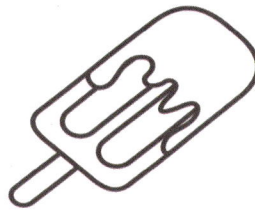

Homemade food is usually better
than factories.

WHY DO WE NEED A CLEAN AND HEALTHY ENVIRONMENT?

Healthy Environment
=
Healthy People

What does being healthy mean to you?..........................

WE AFFECT OUR ENVIRONMENT

Sometimes we make our environment dirty and bad for us. 😔

What do we do to make the environment dirty?...........................

WHAT IS TRASH/WASTE?

When we are done using something and don't need it anymore, we call it trash and throw it away. Some examples of trash are banana peels, paper wrappers, and plastic bags.

A pic taken by Dr. Bee at New Delhi, India, 2023

Can you count 4 waste items you made today?

1................... 2...................

3................... 4...................

HOW MUCH TRASH?

Every year, we make about 2.01 billion tons of solid waste in the world.

In the United States, each person makes about 5 pounds of waste per day or about a tonne every year.

1 kg = 2.2 pounds (lb)

1 ton (USA) = 2000 lb

1 metric tonne = 1000 kg/ 2204.4 lb

How much do you weigh? =

WHAT'S IN THAT TRASH CAN?

Can you look inside the trash can in your house? Look for 4 things that are being thrown away.

············   ············

············   ············

would you like to color this trash can?

WHERE DOES TRASH GO?

Landfills

Rivers

Oceans

Animal tummies

Have you ever seen a landfill?........

WHY IS TRASH BAD?

I don't like trash because -
1. It takes up space
2. It kills animals
3. It spreads disease
4. It adds bad chemicals to our air, soil, food, and water

What is your favorite animal?

....

HOLY COW!

Did you know that the cow is sacred in India? To harm or kill a cow--especially for food--is forbidden by most Hindus. When a cow doesn't give milk anymore, it cannot be killed, and so it is sometimes left to wander the streets.

About 5.2 million cows roam around in the streets of India.

COWS EAT PLASTIC

Unfortunately, nowadays most street cows die of plastic suffocation. This is because
people throw their food waste in plastic bags and cows end up eating plastic bags with the food.

A pic taken by Dr. Bee in New Delhi in September 2023

ARE COWS USEFUL?

Cows give us milk. What 3 foods are make from milk?

1................

2................

3................

Color Me!

Tip - Did you know that cow poop is a very good fertilizer?

GREAT GARBAGE PATCH

Did you know that there is a big patch of garbage in the Pacific Ocean that is twice the size of Texas?

Texas is a very big state in the United States of America!

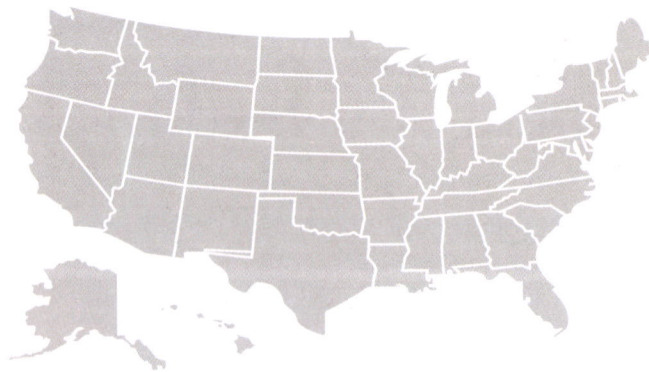

Can you find Texas on the map of the United States? ....

HOW BIG IS THE GARBAGE PATCH?

The Great Garbage Patch covers an approximate surface area of 1.6 million square kilometers – an area three times the size of France and almost half of India (41%).

Can you imagine and color what the garbage patch could look like?

PLASTIC TRASH

Plastic is the main waste item nowadays. Did you know that plastic was not popular until 1950?

Why is plastic a problem?

What do you think were some of the trash items before 1950?

....

COMMON PLASTIC WASTE ITEMS

Shampoo bottles
Plastic cutlery
Styrofoam cups and plates
Packaging
Grocery bags
Plastic straws

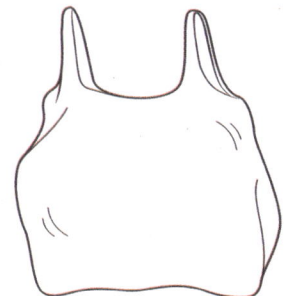

What other items are made of plastic? Name 3.

1................ 2................ 3................

DISPOSABLE PLASTICS

Have you ever used disposable plastic?

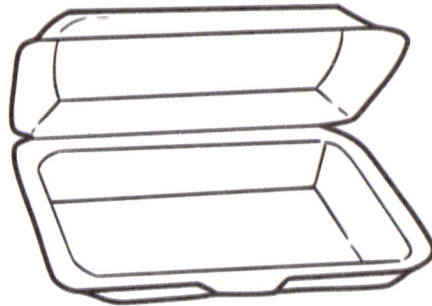

Disposable = Designed to be thrown away after one use

DON'T BLOW THE BALLOON

A lot of people use balloons for decorations and celebrations. But did you know that balloons kill a lot of birds and other animals?

Sea turtles are hit hard as they surface to breathe and eat and commonly eat balloons. Photo by USFWS Eastern Shore of VA and Fisherman Island NWR

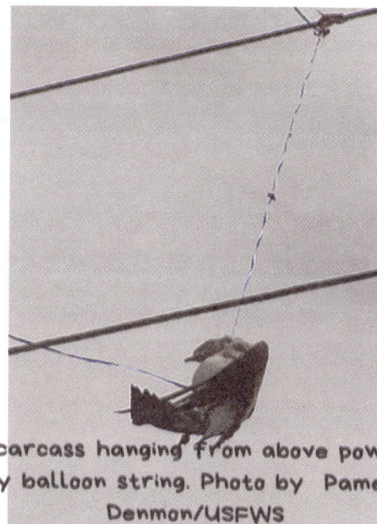

Bird carcass hanging from above powerline by balloon string. Photo by Pamela Denmon/USFWS

Fun fact: Balloons are banned in Disney World's Animal Kingdom theme park and in many national parks in the US.

What can you use to decorate instead of balloons?.....................

HOW MUCH FOOD DO WE WASTE?

According to the United Nations Report in 2021, more than 900 million tonnes of food is wasted every year

Do you remember?

How many kilograms are in a metric tonne?.........

How many pounds are in a ton (US)?.......

WHAT IS POLLUTION?

Pollution happens when the environment is dirtied by waste, chemicals, and other harmful substances.

3 main forms of pollution: air, water, and land

Can you think of one example each of air, water and land pollution?

1.   Air pollution.....................

2.   Water pollution................

3.   Land pollution..................

INDOOR AND OUTDOOR POLLUTION

INDOOR POLLUTION - Inside our house. Example - Burning gas (in stoves or furnaces), Using perfumes, Cleaning with toxic chemicals

OUTDOOR POLLUTION - Outside our house. Example - water pollution from sewage, laundry chemicals, factories

DO YOU REMEMBER CO2?

Do you remember that we breathe in oxygen (O2) from the air to live and exhale carbon dioxide (CO2)?

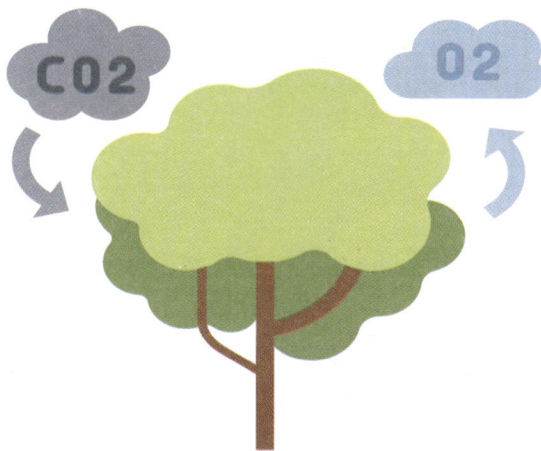

CO2

O2

We are very lucky that trees breathe in carbon dioxide and give out oxygen.

IS CO2 GOOD OR BAD?

Carbon dioxide (CO2) is good for trees but too much CO2 causes pollution.

When we keep cutting trees, and keep making more CO2, we end up with too much CO2!

HOW DO WE MAKE MORE CO2?

We make a lot of CO2 from..
1. Driving
2. Landfills
3. Factories

GREENHOUSE GASES

We make some other gases like nitrous oxide (N2O), and methane (CH4). These are called greenhouse gases.

CO2
N2O
CH4

Carbon dioxide
Nitrous oxide
Methane

What is a greenhouse? Have you ever seen one?

WHAT IS A GREENHOUSE?

Growing vegetables in the greenhouse is possible in cooler climates.

WHAT IS THE GREENHOUSE GAS EFFECT?

Gases like CO_2, CH_4, and N_2O absorb heat from the sun and make the earth's atmosphere hot like a greenhouse.

More of these gases = a hotter atmosphere

Let's revise - Write down the names and symbols of 3 main greenhouse gases...............

....

WHAT IS GLOBAL WARMING?

Our Earth has been heating up due to excess greenhouse gases in the atmosphere.

Where do you live? City..................
Country.....................
What's the average annual temperature in your town?
For the answer, look in your local weather forecast.......................

IS GLOBAL WARMING BAD?

Due to global warming, our earth is heating up and the climate is changing. It can make us very sick.

WHAT IS CLIMATE CHANGE?

Climate change means a change in the average conditions — such as temperature and rainfall — in a region over a long period of time.

Global warming contributes to climate change!

Have you ever heard of climate change?

EFFECTS OF CLIMATE CHANGE

- Heat Waves
- Flooding
- Drought
- Unpredictable rainfall
- Stronger Storms
- Poor Health

What have you heard about Climate Change?

WHAT ARE THE SOLUTIONS?

Looks like we are creating a lot of pollution and creating climate change that is very harmful for us. Do you think we should do something about it?

What do you think we should do?

SOME IDEAS
TO REDUCE POLLUTION!

1.   Eat natural food

2. Use natural products

3. Grow your food

4. Make less waste

5. Compost

What are 3 other ideas you have to reduce pollution?

1..

2..

3..

Feel free to email me your ideas at greenpractitioner@gmail.com.

THE THREE R's

REDUCE

REUSE

RECYCLE

...In that order...

Reduce waste by not creating it in the first place

Reuse materials before recycling or discarding

Transform material into another usable material

Reduce

Reuse

Can you list 3 examples of how you can reduce (Waste)?

1..............

2..............

3..............

WHAT IS RECYCLING?

Recycling means changing waste materials into useful materials.

Different things like paper, glass, carboard, etc. can be recycled instead of thrown away.

Have you ever recycled anything?

OTHER R's

Refuse

Repair

Repurpose

Refurbish

Return

Rot = Compost

Match the "R" with the correct picture

Refuse -

Repair -

Repurpose -

Refurbish -

Return -

Rot -

WHAT IS COMPOSTING?

Composting is recycling food waste into useful fertilizer.

Have you ever tried to compost?

Would you like to color this banana?

Tip - Next time you eat a banana, bury it in a planter... and dig it up a week later!

CAN YOU RECYCLE WATER?

You can recycle water by collecting rainwater instead of letting it go down a sewer.

Question: where does water go when it goes down the drain or into the sewer?

RAINWATER HARVESTING SYSTEM AT RETHINK

In reTHink gardens, we try to use rainwater for growing fruits and vegetabales as much as possible.

WHY PLANT FLOWERS?

Flowers help pollinators like insects, birds, and bats.

What is pollination? - Pollination is very important for plants to make seeds. This is how we get our food.

CAN YOU GROW FOOD?

Farmers grow food. But you can grow some too!

With the help of an adult, find some seeds, and plant them in a pot. Water it and watch it grow. What did you plant? Do you want to send me a picture?

Tip - You might find seeds in your yard or the kitchen!

WHY ARE WE (THE BEES) IMPORTANT?

Bees are important for 1/3 of our food supply! They move pollen from flower to flower.

Color Me, pretty please!

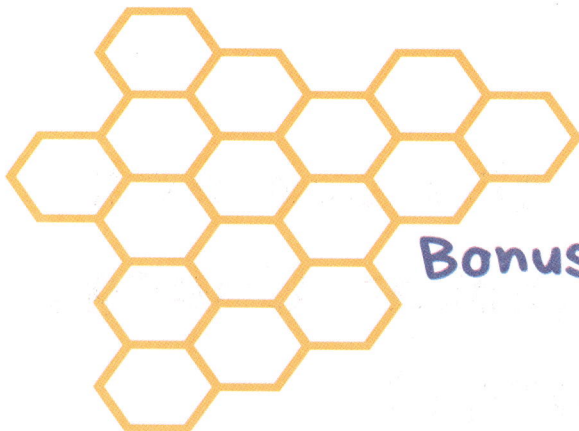

Bonus question - What else do bees give you?

ENERGY MATTERS

We need energy/electricity. But making energy from fossil fuel creates pollution and greenhouse gases.

Bonus question - What is fossil fuel?..............................

But we can also use the sun and wind to make energy (Solar Energy and Wind Energy)

55

SOME PLANTS CLEAN AIR

Did you know that some plants can absorb bad chemicals? Some examples are English Ivy, Snake Plant, and Aloe Vera.

Can you identify and color this plant?

Bonus question - What is one other use of Aloe Vera?...........................

USE NATURAL PRODUCTS

Do you know that a lot of everyday products have harmful chemicals in them that create pollution?

Read the labels of your toothpaste, body lotions, and cleaners. Identify 3 bad chemicals.

1......................

2......................

3......................

EAT NATURAL FOOD

This is one of my gardens in Indiana, USA

If you have a garden, send me a picture of your garden at greenpractitioner@gmail.com. You might be chosen to receive a gift!

WHAT SHOULD WE DO?

> I think we need a clean environment. What do you think?

How will you reduce waste and pollution to create a better environment?
